CLARA FURNESS

How to make
BEESWAX CANDLES

Revised and Enlarged with
CANDLE CUSTOMS and
JUDGING BEESWAX CANDLES

Published by **Northern Bee Books** 2022
ISBN: 978-1-914934-40-7

Tel: 01422 882751
www.northernbeebooks.co.uk

Wax candles have been a feature of religious observance in Church and Monastic life over hundreds of years.

THE RAW MATERIALS

THE expression "wax" almost always applies to beeswax, for, of all waxes, it is historically the most important. The Latin "cera" forms the basis of many of our words relating both to wax and to the modelling for which it was the first raw material.

Even today, beeswax is referred to simply as "wax". If you see "yellow wax" (cera flava) or "white wax" (cera alba) in a recipe, these words apply to the untreated and bleached forms of beeswax.

The delicate scales of newly formed wax are white or translucent. They are sometimes found on the floor of the hive into which a swarm has been shaken. A sudden burst of cold weather; a hive too large for the necessary heat conservation; a frustration of the comb building urge may cause them to fall unused. The swarm, gorged on honey, carries with it the raw material for conversion into wax. Hanging in a living festoon, the bees raise the temperature. Molten wax forms in the eight abdominal glands and solidifies as it comes into contact with air. As it is formed directly from the honey ingested by bees, beeswax can be looked upon as a direct product of nectar of either floral or other source. Before manipulation into honeycomb, the scales are brittle and roughly rounded. One can realise when handling them the origin of those touching stories told about primitive keepers of bees. Finding the graceful constructions in the early part of the year when constricted colonies press upwards with comb building, these simple souls imagined that the bees were creating an altar at which to worship. Their industrious hum was likened to a hymn of praise and the tiny waxen flakes seen as consecrated wafers for the celebration of a miniature Eucharist.

Whilst fashioning the comb, a worker bee introduces into the wax secretions from her mouth, nectar, honey, brood food remains and soiling from her feet which "handle" it. Pollen enters it and contributes a wide range of colouration. Existing soiled wax is extended and mixed with the new. This is apparent if we compare the colour of brood or honey cappings on new and old combs.

By the time we come to render the wax for our own use, its colour is affected by all these materials. We want the brightest and cleanest if we are hoping to enter the wax, and candle classes at a honey show. So it is as well when collecting wax to separate it into two grades.

Even the best managers of bees will sometimes have free comb built in their hives. Do not regret this, but add the clean wax to your cappings which supply most of your first grade. When extracting, allow only cappings to fall into the tray. Drop the frame scrapings into another receptacle. Be frugal with scraps of burr and brace comb, keeping a collecting box handy during all manipulations. Regular comb replacement for combatting Nosema disease gives good wax yields. Try to keep this second grade as clean as possible, though do not hesitate to pick up the scraps thrown down by less frugal colleagues.

A solar wax extractor is essential to a beekeeper who has become interested in candle making. It is surprising what golden wax will flow from discarded combs and discoloured scrapings. Sunshine was the chief bleacher for wax in ancient days. Your county probably has an area still called "Bleaching Fields" or some such name. This recalls the time when beeswax was shredded and remelted in summer sunshine several times to lighten its colour.

After rough extraction the wax needs to be filtered to remove dross and dirt held in suspension. It melts at 62°-64°C (144°-148°F) and will discolour if heated above 85°C (185°F).

Several useful articles have been printed about the production of cakes of wax for competition and these apply even more to wax for candles. The lovely natural colour of beeswax is to me far more attractive than bleached wax, but some of its colour probably comes from the propolis incorporated in it. It is said that bees coat even new comb with a fine layer of propolis and this will melt into your wax. Scrapings will contain a great deal. Unless the fine grains of propolis are excluded from wax intended for candles, the candle will not burn properly. As the wax melts and feeds the candle flame any propolis in it chars. The wax becomes flecked and discoloured and the charred particles clog the wick and cause the flame to crackle and splutter.

My own filtration method is to heat my electric oven to a temperature of about 70°C (160°F). [Use a thermometer and then scratch a mark on the control switch.]

I collect large food cans from a catering establishment and clean them thoroughly. These conveniently support a metal sieve over which I lay a piece of clean cotton material or strong paper towel. I melt the wax in a jar standing in a pan of water, and then pour it into the sieve, leaving it in the oven to drip through. This sieved wax looks clean, but if it is now passed through a filter paper* folded into a metal funnel, (plastics soften at this heat) a surprising amount of very fine soiling remains on the paper.

All these processes are inefficient of course and large amounts of "slum glum" are collected in the early stages. Absorption by filter cloth and filter paper also represents unfortunate waste. Frugal beekeepers who have open fires in their homes can make good firelighters from the waste. Mix with sawdust, wood shavings or just roll small quantities in sheets of newspaper. Small, but significant returns of wax can be taken from this waste by our commercial wax men who have efficient techniques

* Available on amateur wine counter at Boots or other stores.

for extraction. In 1973 beeswax was imported at £800 to £900 a ton. The price is now treble that and we can imagine the value of our home produced stuff.

The quicker we can render our beeswax and the less we expose it in warmed air, the better we can preserve that rich aroma redolent of all the flowers of its originating nectar. Our English beeswax has the delicate scent of our temperate landscape, but I remember a visit to the wax importers, Poth-Hille and Co., at Stratford, London E.15 where the office, full of samples of wax from flamboyant tropical lands, smelt like a hothouse at Kew.

The knowledgeable men who deal with wax from all over the world, have sophisticated techniques for checking quality and detecting adulteration. Their extraction methods are thorough and their bleached wax is white as snow. Their practised senses can tell them the probable life history of a piece of beeswax when they examine the texture of a break and smell the released fragrance. Mr. Case-Green of British Wax Importers, Redhill, Surrey may be seen demonstrating this expertise when he judges (in public) the commercial wax class at the London National Honey Show each year.

SAFETY

DISASTROUS fires have occurred in bee sheds owing to accidents with wax rendering. The very reason we are making beeswax candles is because here we have a material which will vaporise and burn with a bright, steady flame. It is essential before embarking on candle making to be aware of the dangers and well equipped to deal with any mishap which may occur.

Electricity is the best source of heat just because it does not involve a flame. The vessel holding the wax should stand in a larger vessel containing hot water. It is never necessary to boil the water, as you will never need wax as hot as 100°C (212°F). Even so, containers, handles, etc. will become hot and jumpy fingers cause spills. So keep kettle holders or oven mitts handy or best of all, learn to work wearing lined rubber gloves.

All candle making (or wax modelling) requires patience. Wax melts slowly and does not heat evenly. Desired changes of temperature must be waited for. You can't do a rush job by popping the melting wax over direct heat. Beeswax begins to discolour as soon as it reaches 85°C (185°F). Stand a thermometer in the water bath and do not allow it to exceed this temperature.

It is always safest to empty your wax pot

after each work session. Pour the remaining wax into shallow polythene containers from which it is easily released. Break it into pieces for re-melting. If you do leave wax to set in your melting pot, it is essential that the surrounding water comes as high as the surface of the wax. The wax will then melt all round and release itself from the sides of the pot. If the wax melts from underneath, molten wax will expand and press against the solid surface wax adhering to the sides of the container. A messy explosion of hot wax can occur when this pressure suddenly bursts the surface layer.

When dipping, allow for displacement or wax will spill over into the water.

When pouring molten wax, see that the vessel has a proper handle and pouring lip. Have moulds and pouring areas well away from the source of heat.

The would-be chandler should look at a few pictures of old fashioned candle manufacture. Wax is festooned all over machinery, counters and candle maker. It is these old techniques we amateurs must emulate. The clean, mechanised, modern candle factory does not use melted wax. Instead, wax is powdered, forced into a suitably shaped chamber having a wick threaded up the centre, and extruded under pressure. A few still employ molten wax for limited production of liturgical candles, but the number is dwindling.

All floors and surfaces should be protected. Newspaper is not a good covering for surfaces near the stove. It can easily flap and catch fire. It is, of course, adequate for the floor. Plastic coated hardboard is an excellent surface and offcuts can be obtained cheaply. This can be cleaned and used over and over again. If you should have a spill, wait until the wax is just setting, when it can be peeled off most surfaces and returned to the pot. If it is allowed to go quite cold, it must be chipped off with a hive tool and the surface may be scratched, or worse. Warm wax on hands is just a beauty treatment. Peel if off when partly set. Hot wax can cause a burn. Don't try to wipe it off, dip the hand in cold water or run the tap over the wax and peel it off. Slap honey over the affected parts and they will heal with no scar.

Try to stay with your candle making once you start. If called away, turn off all sources of heat. Don't take risks. If, in spite of all care, you do start a fire, there should be an extinguisher near your work surface. Water is of no use for extinguishing burning wax. The ignited material just floats on the water which splatters and spreads the fire. Buckets of sand and large wet cloths should be at hand if you have no extinguisher.

If things seem to be getting out of hand, don't hesitate to call the fire service.

HISTORY

BEESWAX has been an object of wonder and worship as long as mankind has had use of it. A. L. Cribb, writing in *Bee Craft* in December, 1944 said:—

"In Chaldee, Dabar signified the word, but it also meant a Bee. so the bee was used as a hieroglyph. Wax was considered to come out of the mouths of bees and so was revered and used in the manufacture of tapers and candles in Pagan worship. The light of the wax candle of Dabar the bee was set up on pagan altars as a substitute for the light of Dabar the word . . . that enlightened the souls of men."

Wax played its part in Greek and Roman mythology. Every schoolchild learns the story of Icarus who found beeswax to have too low a melting point for Greek sunshine when he used it to attach his wings for his abortive attempt to fly from Crete.

There is much discussion in Bible studies on the use of the Hebrew word n'er, translated variously as "lamp" and "candle". It seems probable that light was produced by whatever combustible was most accessible to early peoples; and it is likely that lamps of olive oil with floating wicks preceded candles in Mediterranean countries. Candle materials would, indeed, melt or at least bend in these very hot places.

Christ referred to himself as "The Way, the Truth and the Light". In the first three centuries of Christian worship, spiritual simplicity and antagonism to heathen observances were stressed. Tertullian, about 200 AD, explained that the Heathen gods, who are of the earth, have need of light, but the Christian God is Himself the giver of light, even of the Sun and therefore has no need of candles in His honour. This observation leads us to believe that lamps and candles had been a strong feature of Greek and Roman pantheism.

St. Paul himself advised the early evangelists not to reject pagan worship out of hand, but to absorb what was familiar and secure into Christian observance. The offering of precious materials such as beeswax and the obvious symbolism of light soon assured the introduction of wax candles.

As influential people, such as the Emperor Constantine and King Ethelbert, were converted, many old customs returned. The ceremonial use of lights was of special significance, particularly at Easter time and was the symbol of joy as the faith expanded. Flames were carried before the Pope as a mark of honour and before minor clergy when new churches were dedicated.

There is no firm record of candles on the altar as a part of the required observance until 1175 when two candles are described as "the present custom of the papal chapel". As time went on, however, more candles were used for important festivals. They were also bought as offerings before the images and shrines of Saints with prayers for favours, recovery or other benefits.

It was natural that a special festival for candles developed. This, like so many Christian feasts, was built upon the foundation of earlier beliefs. Northern peoples welcomed the returning Sun with fires and torches. The Romans worshipped Ceres, whose daughter, Proserpina, was stolen by Pluto and taken to the Underworld. This represented Winter and as Ceres sought her daughter with torches and candles (which she was said to light in Mount Etna), a sacrifice of lamps and candles was made for the return which represented Springtime. The Latin word Februo (I purify by sacrifice) remains in our word February; and it is on the second day of this month that the early Church placed Candlemas. Superimposed on the primitive thank offerings for the sun's return, was the dual celebration of the Purification of the Blessed Virgin and the Presentation of Christ in the Temple. The year's supply of beeswax candles were blessed by the clergy and carried in procession, while the priests sang the *Nunc Dimittis* and the people were reminded of Simeon's words: "A light to lighten the Gentiles".

The prayer for the consecration of the candles was "Lord Jesus Christ, Son of the Living God, Thou true light which lightens every man who cometh into the world, we pray Thee to bless these candles, that, wherever they are lighted, our hearts, enlightened by the invisible fire and purity of the Holy Ghost, may be freed from all blindness of sin and vice, and that, after the dark and dangerous pilgrimage of earth, we may enter into everlasting light". The simple folk, still held by superstition and fear, often kept the remains of these candles as a sure protection against the things they most feared—personal illness, diseases of cattle, failure of crops and stormy weather.

It is easy, when we see the pure flames of our own beeswax candles to appreciate the power of the symbolism of these lights in the dark days of old. They were not just to ensure that the priest saw the celebration of his office in the dark chancel, but represented that natural seeking of the human spirit for a realistic representation of truth and revelation.

Growing theology elaborated old custom and naturally enjoined that church candles should be made only from pure beeswax. The sole alternative was tallow—the drippings from the cooking of fat meat. These candles burned with an acrid, smoking flame and were not suitable for the homes of the nobility or the altars of churches. Because the bee was never seen to mate like other animals, it was extolled as an example of purity and virginity. "The fragrant wax, the labour of the bee, which dies when its work is accomplished, has mystic significance. It is drawn from the best juices of plants and has the highest natural worth as a material for offerings."*

Of course customs became institutionalised and observance complicated. Candles of pure, bleached beeswax were made obligatory for all liturgical services and ceremonies. The Worshipful Company of Wax Chandlers was formed sometime in the 14th century and incorporated by royal charter in 1484. A copy of this charter with its huge seal of beeswax hangs in the "greate room" of the Wax Chandlers' Hall in the City of London. This craft guild was originally a trade association of self-employed craftsmen, combining to protect their own interests and standards in an age of doubtful honesty. They took upon themselves the authority to inspect anyone using "the art and mystery of Wax Chandler". They were "to discover abuses, deceits and falsities in work-

---

* *Encyclopaedia of Religion and Ethics.*

manship and materials tending to the wrong and detriment of His Majesty's subjects and to the scandal of the said art". The 13th and 14th centuries shew records of the use of much beeswax in England, most of it for great occasions enjoyed by the nobility and high churchmen. When the reformation came, this was scorned as "a time of superstition" and all wax candles, tapers and images were prohibited in churches throughout the land. Commissioners were sent to "take away, utterly extinct and destroy all candlesticks, tyndrilles or rolls of wax", and the clergy were ordered to preach against the superstitious offering of candles and tapers to images and relics. This must have been a lean time for the Worshipful Company of Wax Chandlers. But candles came back at the restoration with 100% purity of beeswax required for all liturgical lights. These were of bleached wax, as, indeed, they still are. For requiem masses, funerals and the Good Friday liturgy, however, the candles were to be of unbleached or yellow wax.

Candles continued through several centuries to be used in much the same way. Church candles gained new significance in the nineteenth century with strong religious controversies and Catholic emancipation.

Throughout this long period, candle making for the Church and noble houses was still a craft, but homely tallow lights were made by thrifty housewives and their servants. In "Scenes from Clerical Life", George Eliot describes a tallow dip as "an excellent thing in the kitchen candlestick and Betty's nose and eye are not sensitive to the difference between it and the finest wax". When the Countess Czerlaski is expecting visitors, she contents herself with the light given off by the fire, though "on the table are two wax candles which will be lighted as soon as the expected knock is heard at the door."

Soon, however, new materials were discovered which proved to be cheaper than beeswax. The development of science led to discoveries about the nature of fats, waxes and oils. Spermaceti (whale oil), stearine (from animal fats after the removal of glycerine), coconut oil, palm oil and, finally, paraffin were introduced for candles.

The use of paraffin wax for candle making dates from 1850 when James Young produced it from crude petroleum found in an oil spring in Derbyshire. Later it was extracted from bog-head coal mined in West Lothian, Scotland. The real development of separation of oils and waxes from Burma oil started in 1853 and expanded rapidly throughout the world. This was the onset of modern candle manufacture. These materials were found to be acceptable for church use and, little by little, admixtures with beeswax were permitted. From 1850, beeswax had to form only "a notable part" of candles lighted during mass and this was usually interpreted as 95%, though candles of 80%, 75% and even 65% were accepted.

Recent years have seen a huge increase in the price of beeswax and, after the second war, candles for liturgical use (all large Easter or Paschal candles and altar candles for mass) were required to have "the greater part" of beeswax. This led to a church candle containing 51% beeswax and the rest paraffin wax and stearic acid. Other church candles could contain only 25% beeswax. In 1975, this proportion was accepted for the special candles and ordinary, extruded paraffin wax candles are used elsewhere in churches. When I visited Price's Candle factory in 1975, where liturgical candles are still made by hand, their craftsman was endeavouring, with little pleasure, to produce the traditional quality candle with these new proportions: 25% beeswax, 65% paraffin wax and 10% stearic acid. Of course, the recent success with the separation of a vast number of grades of paraffin wax has given the candle industry waxes with the same pure, smokeless flame as was the exclusive prerogative of unadulterated beeswax in days gone by. The making of pure beeswax candles is now a luxury confined to those eccentric people who keep bees.

It is sad that the dark centuries long past did not have such lovely candles and that we were enjoying electric illumination when they were developed. It will also be sad when the simple symbolism and mysticism of early Church has quite gone and such touching words as those used by John Dunmelow in his recent history of the Wax Chandlers of London are no longer relevant: "In Christian worship, the candle typifies Jesus Christ, light of the world: the wax represents the spotless body (the bee being looked upon as the virgin producer of offspring); the wick His soul, and the flame His divinity". I have great faith that this sense of wonder is not lost. It is a far cry from the day when monks, lost in the contemplation of a virgin birth and the apparently spontaneous generation of bees, painted them as being supplied to eager mankind directly from the hand of God. Yet, new recruits to beekeeping continue to enjoy a mystic enthusiasm even when bombarded with all the modern, scientific knowledge.

A romantic spell

In spite of the rapid succession of oil lamps, gas light and electric illumination, candles still spell romance. Whenever there is a festival—on birthday cakes, at Christmas, with a celebration meal, for Easter day, the candle still shines. Candle making has become a popular hobby and books describing the craft are numerous. Unfortunately each one deals only with the modern, versatile, cheap paraffin waxes. They dispense with beeswax in one short sentence—it is too expensive—and so they will not give any advice on how to use it.

The candle classes at the London National Honey Show must be the last relic of a craft once so widespread and profitable. One has only to see the unique individuality of each entry to appreciate that each candlemaker has had to learn from first principles for himself. Over the years, a few have developed some technology by trial and error. I have, however, been unable to find one publication which fully described anyone's techniques and methods. My own experience is very limited. My early enthusiasm stems from a lecture-demonstration given by the late Miss D. V. Burch in 1971. My apparatus has been home-made, crude and inadequate. I wish I had some skill in wood and metal work to extend my equipment. I have, however, produced a small number of candles which have given me great pleasure. In a busy life, where very little time could be devoted to this small branch of my hobby of beekeeping, I have experimented with wax and wick, time and temperature, shape and design. Miss Burch answered my letters with patience and generosity and I owe most of my "know-how" to her.

I know that the only way to learn about one's own candles is to burn them. Unfortunately I still feel guilty when I set light to this exquisite, fragrant, golden product of my bees. Perhaps a few more beekeepers will make candles and raise the standard at the N.H.S. I would be glad if they would send me news of discoveries they have made, so that beeswax candle making may be re-learned and recorded for future beekeepers.

A first century candle unearthed at Vaison, France (see page 6)
Photo: British Museum

THE WICK

THE golden candlesticks of the Bible and those of bronze and gold found in the treasures of Tutenkamun's tomb were not for candles as we know them. They were to support receptacles for oil in which a wick floated. The first "candles" were torches made by dipping whatever was available in fats, oils or waxes. The difference between a torch and a candle is in the relation of wick to combustible. In a candle the material surrounding the wick is much thicker than the wick itself. In a torch the wick and the combustible are about the same size and burn away at one and the same time.

There is historical evidence that candle making with beeswax is of ancient origin, but wicks have changed. A lamp and candlemaker's shop was discovered in the ruins of Herculaneum destroyed by the eruption of Vesuvius in 78 A.D. A first century candle with a diameter of 12.1 cm (about 4¾ in.) was unearthed at Vaison in France (see illustration). An Alemannic grave excavated in Germany revealed a large candle with a wick consisting of 9-10 woven threads. Viking remains on the German coast show that candles 40 cm (16 ins.) long and with diameters of 8-10 cm (3¼ to 4 ins.) were used. None of these candles was used, of course, by common people. They were for very special occasions. Most ancient peoples looked upon their rulers as sharing a divine origin, so religious observance and royal festivity were often combined.

There is evidence of how Viking candles were made. A cake of wax one to two cm. thick was kept warm till it was almost pliable. From this, strips about 2 cm wide were cut. A linen wick was placed between two such slabs and the block was rolled till a round candle was made. Four strips of square section could also be put round a wick and rolled in the same way. I have seen beautiful candles at the National Honey Show still made by this ancient method.

The wick, often referred to as "the soul of the candle", is that part which carries the molten wax upwards by capillary action to where, in contact with air, it vapourises and is burned. The flame feeds downwards along the wick, but is stopped by the fluid wax. The wax must become gaseous before it can burn. Under the flame, a shallow dip develops to hold the wax as it melts. The outer edge of the candle is kept cool by the rising air which also draws the flame upwards and feeds it with oxygen. The dark space in the middle of the flame holds the gases issuing along the wick, which, not being yet in contact with air, cannot undergo combustion. Small deposits of carbon appear on the end of the wick, but, as this bends over and meets the hot, outer edge of the flame, this burns away.

Going back to the earliest wicks, no doubt our primitive ancestors used burning brands from the fire for lighting the cave. They soon realised that those from resinous, coniferous trees burned longer, while those used to poke the fatty meat burned longer still. These burning sticks, especially those of straight-grained pine, rich in resin, were used by all Northern peoples. Splinter candles survived in remote Russian villages into the last century.

Torches or flambeau daubed with resins, fat, oil or wax were a natural development. Strings of flax and wool shearings were drawn through these materials in ancient Greece and Rome. Egyptians used the plentiful reeds and dipped them in beeswax, whilst Eskimo peoples made use of animal tissue in pans of seal blubber. In England the absorbent pith or core of rushes and sedges were used and the ingenious candlemaker discovered that if a strip of the outer covering was left on one side, the wick would be drawn into a curve and consumed in the hottest part of the flame, thus minimising smoke. Common tallow candles, whatever wick was used, gave off acrid fumes and dense smoke, but had the advantage of being edible. They are said to have preserved the lives of many marooned explorers and lighthouse keepers when all other rations had been consumed.

It is not known when cotton wick was introduced, but it probably replaced those made of linen as cotton production developed.

At first wicks were made of twisted threads and the tufted material "candlewick" (now usually used for bedspreads) is an example. Hand-dipped candles may still be bought in tourist markets in the Greek islands, made with twisted wicks. These wicks remain upright during burning, and the end of the wick stays in the cool interior of the flame. A head of black soot, called in the trade "a cauliflower", forms on this end, dims the illumination and causes a great deal of dark smoke, unless the offensive end is trimmed away.

Early candles had to be trimmed or snuffed at frequent intervals. Small scissors equipped with little boxes to receive the cut end were used. These candle snuffers, now collectors' items, were the bane of life right up to the last century. Children and servants had to go round every half hour or so to trim the candles or the air became full of smoke. The German dramatist and poet Goethe is said to have remarked how much more could have been achieved and discovered by man if only candles did not have to be trimmed. Restoration comedy was played in brightly illuminated theatres, but the action was interrupted very frequently whilst armies of attendants snuffed the candles. Otherwise the actors would have become almost invisible or the audience asphyxiated by the fumes.

Using the peeled rush of the rush light as inspiration, in 1825 a Frenchman, Cambacérès discovered that braiding or plaiting the wick caused it to bend over whilst burning. The carbonised and incandescent end thus came into contact with the hot, outer part of the flame and could utilise the oxygen in the air and be totally consumed.

All the wicks available to us as amateur candlemakers are braided. Cotton is the material employed, as this is naturally absorbent. Till recently it was possible to plait one's own wick with darning cotton, but this arduous task is no longer necessary as ready-plaited wicks are in abundant supply. Beekeepers are advised not to try to be total craftsmen by plaiting their own wicks, as most "cotton" threads now contain an admixture of man-made fibres which are low absorbent and do not burn properly. Since the cotton for wicks must be well woven, smooth and have no loose ends, it is as well to leave the job to the professional. Braided wick is sold by numbers and the European metric cotton number (Nm) indicates the number of metres of yarn which weigh 1 gram.

For example, 4 metres weigh 1g. 24/1 is the Nm which applies to 1 thread. 24/2 is the Nm which applies to 2 threads.†

Flat wicks or plaited wicks are made from three bundles of the same thread number plaited or braided like a girl's hair.

If the wick is plaited too loose, it sags and sinks itself in the bed of melted wax. If plaited too tight, the absorption power of the wick is restricted. Therefore, not only the thickness of the wick, but the tension of the plaiting governs its performance.

Wicks for sale are described by the yarn number and the number of threads in each section of the braid. Thus 20/3X9 indicates three bundles of nine threads, each thread of Nm 20/1. You will find this specification in the books on Candle Craft, and they will be recommended for certain diameter candles:

e.g., up to 7mm—flat wick of 20/9
(20/3X3)
up to 15mm—flat wick of 20/18
20/3X6)
up to 25mm—flat wick of 20/27
(20/3X9) or sometimes 3X9(20).

† *The English numeration (n) is based on the number of hanks containing 840 yards (768m) to the pound (453.6g). No doubt this disappears as metrication advances.*

This should never be followed for beeswax. It applies to candle wax which is usually 90% paraffin wax and 10% Stearin (or stearic acid). Beeswax has a much higher viscosity than paraffin wax, so a higher wick count is required. Thus a balance between the molten beeswax and the necessary absorption efficiency of the wick is reached. In fact the addition of as much as 5% to 10% beeswax to paraffin wax makes a much thicker wick necessary. All my first candles were seriously under-wicked because I had only the books on paraffin/stearin candles for reference. Pure beeswax candles need at least double the wick recommended for ordinary candle wax.

There are other wicks, plaited with eight to eleven groups of threads and called round wicks, though they seem to me to have a triangular section. They are symmetrically plaited bundles of cotton yarn in which two or three plaits are bonded by division of one group to bind the whole structure together. In Germany they are recommended for all wax mixtures with a beeswax content of over 5%. These round wicks are said to have greater stability than flat ones and to be essential for all very large candles. Round wicks have a "head"— that is they must be used only one way up. This is where the forking of the braid on the flatter side looks to run upwards. The strands stand open and the points of contact go downwards. When you buy these wicks, you find a knot at one end and this is the end that must go uppermost in the candle. I have very limited experience of this wick which is not available in Britain, but on sale in Germany and the U.S.A.

A further factor affects the choice of wick. For perfect combustion, the wicks are pickled in a dilute solution of mineral salts for about 24 hours. This helps to increase illumination, decrease smoking and after-glow and help the bending of the wick to a perfect angle for total combustion.

After steeping long enough for thorough impregnation of the fibres, the hanks of wick are placed in a centrifuge to expel the solution, and then dried. The ability of a wick to take up the molten wax (capillarity) must be such that a steady, clear flame burns, but not quite all the wax which melts is drawn up and consumed. A pool of molten wax remains for the wick to draw upon. If the wick is too thin an excess of wax is melted and the wick cannot absorb it. The excess runs or gutters down the candle. If the wick is too thick, there is insufficient molten wax for it to draw upon and the candle flickers and gives off smoke.

When using your wick, it can be stretched or left quite slack. This will, of course, affect the performance of the candle. The only real path to success lies in making experimental candles and burning them for at least an hour and then adjusting the balance of wax and wick.

A wick with a metal core is mentioned in many candle craft books. Thick candles are moulded and a hole drilled along the centre.

The stiffened wick is threaded through and secured with molten wax. This wick is not available in England because of air pollution caused when the wick burns.

Although pure beeswax candles are not made commercially today, a few candle factories do honour isolated special orders. For this purpose the manufacturers of wick still make a small quantity of wick especially pickled for beeswax. Look up Church Candle Manufacturers in your "Yellow Pages" and ask them to sell you a roll of wick whenever they are making one for beeswax.

FOUNDATION CANDLES

YOUR FIRST CANDLE

THE simplest beeswax candle is one of which is made from a sheet of unwired foundation. Dyed beeswax is made into "foundation" specially for candle making, but I much prefer the colour of natural wax.

As all beekeepers know, foundation tends to be brittle when cold and so a warm room is necessary.

Place the foundation on a clean, warm, smooth surface—Cut a wick a little longer than the foundation and lay it about 5mm from the edge of the sheet. Gently turn up the edge to cover the wick—Then roll the whole sheet around the wick, taking care to keep it smooth, straight and tight. Secure the edge by holding it for a few seconds by a warm surface such as a radiator and then stick it down with gentle pressure of your fingers. Tall, thin candles result from rolling from the longer edges. Small, squat candles from half sheets.

Tapering candles are made by cutting a sheet of foundation in this way:

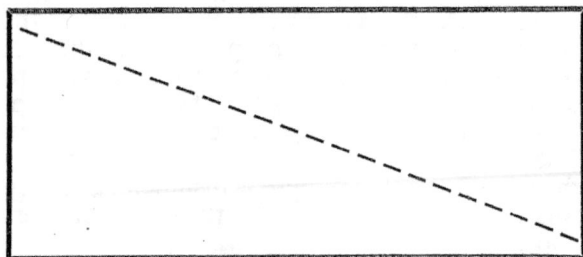

Cut foundation with sharp scissors into petals, leaves or geometric shapes. Thin strips can be woven on the outer layer to look like basket work. You can have fun with design and make candles in all shapes and sizes.

A foundation candle burns well owing to the air trapped inside the structure.

A great improvement is effected by dipping the wick in molten beeswax before starting to roll the foundation and it adheres to the edge of the sheet very readily and rolling is much easier. The candle ignites more promptly and burns longer and more steadily.

POURED CANDLES

When I visited Price's Patent Candle Factory in London in 1978 I saw this traditional method in use. Unfortunately, along with other candle manufacturers, Price's have had to abandon this time honoured technique. The materials permitted by Church edict, 25% beeswax, 55% paraffin wax and 20% stearin lend themselves to extrusion methods. At the same time the work is arduous and recruitment of apprentices difficult.

Candle wicks are suspended at intervals from a circular hoop and a ladle of molten wax is poured down each wick so that a thin film adheres to it. The process is repeated until the candle is the desired size. The candle-maker at Battersea was highly skilled and his long, even candles were masterpieces. To keep them straight and smooth, they are rolled by the hands on a marble slab. Sixteen candles are made on one hoop.

For the amateur this is a simple process, requiring little apparatus, but a maximum of skill. It permits the making of a candle of any desired size. The enormous Paschal candles, famous in past ages, for the Easter mass were poured by beekeeper-monks and often were as tall as the man who made them. The disadvantages of this method are that it is messy and can lead to spills and splashes and it demands large quantities of wax.

You will need: A large pot of molten wax, ladles with wide pouring lips, suitable wick, and some support for hanging the candle. Decide how many candles you are going to make and cut wicks about 15cm (6 in.) longer than the finished candle.

A cool room is an advantage so that each layer of wax sets well. Do not, however, open doors or windows to cool things down, as there should be no draughts. Even areas of different temperature will cause the candle to curve towards the cooler side where setting is slightly quicker. If you work near a window, you will see the candle curve in that direction.

Heat the wax to 70°—75°C (160°—170°F). Immerse the wick in the molten wax, moving it around a little, so that it becomes completely impregnated. Bubbles will rise from it as wax replaces the air trapped between the threads. Draw the wick out straight, but try not to stretch it. Clip the end in a spring clothes peg and hang it up to cool. Repeat the dipping, but this time only for a second or two, so that more wax adheres and the first wax is not melted off. Allow to cool again.

Now lay the wick on a smooth surface. A piece of marble or glass is ideal, but a plastic work surface will suffice. Using a sheet of glass or a smooth bit of wood or laminate large enough to deal with the whole candle at once (excluding the excess wick), roll the waxed wick backwards and forwards to make it rounded and smooth. If you can still see the marks of the plait, repeat the dipping and rolling.

If the baby candle is too warm, wax will stick to the glass. If it is too cold, the candle will break and no amount of effort will mend it. You will have to start again or make two short candles. To be quite sure, keep a large bowl of water at 27°C (80°F) and dip the candle in this. Wipe with a small cloth and then roll it. Completely soft water with a dilute solution of pure soap is helpful. If no large vessel is available wipe your rolling surface very sparingly with the soapy water.

Now suspend your wicks and, holding them in turn over the wide-mouthed melting pot, give the free wick at the top a turn and pour the wax down the candle as it spins to return to its free hanging position. Watch the candle grow even and round. If you see any distortion in the shape, remove the candle from its clip and roll it on your slab. Use your hands now for rolling. Do this gently and lovingly. It not only shapes your candle to a satisfying roundness, but is a beauty treatment for the skin of your hands. You are less likely to break the candle at this stage by rolling it with your hands than if you use the glass (or other) sheet.

An icicle of wax will form at the bottom of your candle beyond the wick. This may be cut off at intervals and the pieces put into the supplementary melting pot for filling up. Do not put these lumps where your ladle may pick them up or they will cling to your growing candle and cannot be smoothed off.

If you are making a very heavy candle, soap a blunt knife and cut the bottom off the candle, leaving the wick intact. Tie a knot in the wick to prevent the candle slipping right off its wick.

If your candle becomes uneven or if you wish to model shapes on its surface, you can make it malleable by submerging it in a bath of water at 49°C (120°F) for an hour or more.

When you are satisfied with the size and shape of your candle, allow it to cool completely. Then boil a kettle or pan of water and stand it on a sheet of aluminium foil. You can shape the top of your candle by rolling it against the hot kettle and flatten the base in a similar way. The drops of wax will crack easily off the foil and can be returned to the melting pot.

NOTE. In a poured candle, the bottom of the suspended candle is the top or burning end of the finished article. At this stage a poured candle is slightly narrower at its base.

If you wish to use a candlestick with a pricket, make a suitable hole in the base of the candle with a hot awl or skewer.

Trim the wick to about 1cm (½ inch) and allow your candle to mature for 24 hours before packing it away.

DIPPED CANDLES

At first it was the tallow candles that were dipped. In fact, they were often just referred to as "dips". Beeswax candles were poured.

From my limited experience, however, I would say that dipping is the best technique for the amateur-beekeeper-turned-candlemaker. In fact, the professional candles described in the last section were finished off by dipping. The pouring process made them slightly narrower at the base (i.e., the *top* of the pour). To render them cylindrical, they are dipped a quarter of the way up, then one third, then a half. When the candles are thus "bottomed up", their size is checked with a pair of callipers. A final full dip in hotter wax, gives them an overall smooth finish for a final rolling.

Root* warns us that beeswax alone is NOT a good dipping wax. When melted it is thicker than much less expensive substitutes such as paraffin and stearin.

We have to disregard this denigration of our chosen medium and realise that we are using the real aristocrat of candle material and that dipping is our obvious technique. Our supplies of wax are limited. We do not need vast quantities of spare wax if we are willing to dip one candle at a time.

There is no proscribed apparatus and mine is singularly home-made in a "Heath Robinson" manner. My water bath is a tall can made by soldering together two catering size food tins. The dipping pot is a tall cylinder made with three frozen fruit juice cans. Wax is lighter than water and the wax pot must stand firmly erect in the water bath. You cannot do with it floating away from the base of the outer can and catching on the candle. I therefore set mine into a honey jar lid filled with molten lead from a bit of old lead piping. [A fellow bee-keeper later made me a superior model from a

---

* *Root, A. I., "Beeswax".*

piece of steel piping by welding on a heavy base]. A beautiful dipper made from a piece of an obsolete Victorian telescope proved to be of brass and discoloured the beeswax.

The illustration shows a laboratory measuring cylinder made from pyrex glass – a perfect dipping pot. Unfortunately it was broken and replacement was very expensive. They are now made in good quality heat resistant plastic and may be bought from suppliers of scientific equipment.

The candles must hang between dips and I have a simple home-made rack consisting of a wooden bar with six nails at regular intervals. This I support on a retort stand so that the nails slope slightly upwards and the candles won't drop off. The wick is suspended from a wooden spring clothes peg. The peg may be hung on the nails either by slipping the hole in the spring on to the nail or by hanging it with a loop of wire through this hole. (String makes the candles sway too freely). This apparatus was designed by my husband when the wire coat hanger I was using tended to swivel on its support. The coat hanger is, however, satisfactory if it can be made rigid on its support.

Otherwise it pivots as the candles are removed and replaced.

Recently I made some large, heavy candles for a special Church ceremony. The spring clothes peg had insufficient grip to support the weight, and the candle dropped off. In this case I had to find something with a firmer grip. I used a metal spring clip with success. But the clothes peg is suitable for all except jumbo candles.

When planning your apparatus remember that the finished candle will be 3-5 cm (1-2 inches) shorter than the dipping pot owing to:

1. The icicle, or "port", which forms at the bottom.
2. displacement of wax as the candle grows thicker.

If your dipping pot is too near the size of your desired candle, i.e., too narrow, you have to take great care not to touch the sides as you dip the candle. A much larger dipping pot makes for easier candle making, but in this case you need lots of wax to fill it. You have to strike a balance between ease of working and your supplies of cleaned beeswax.

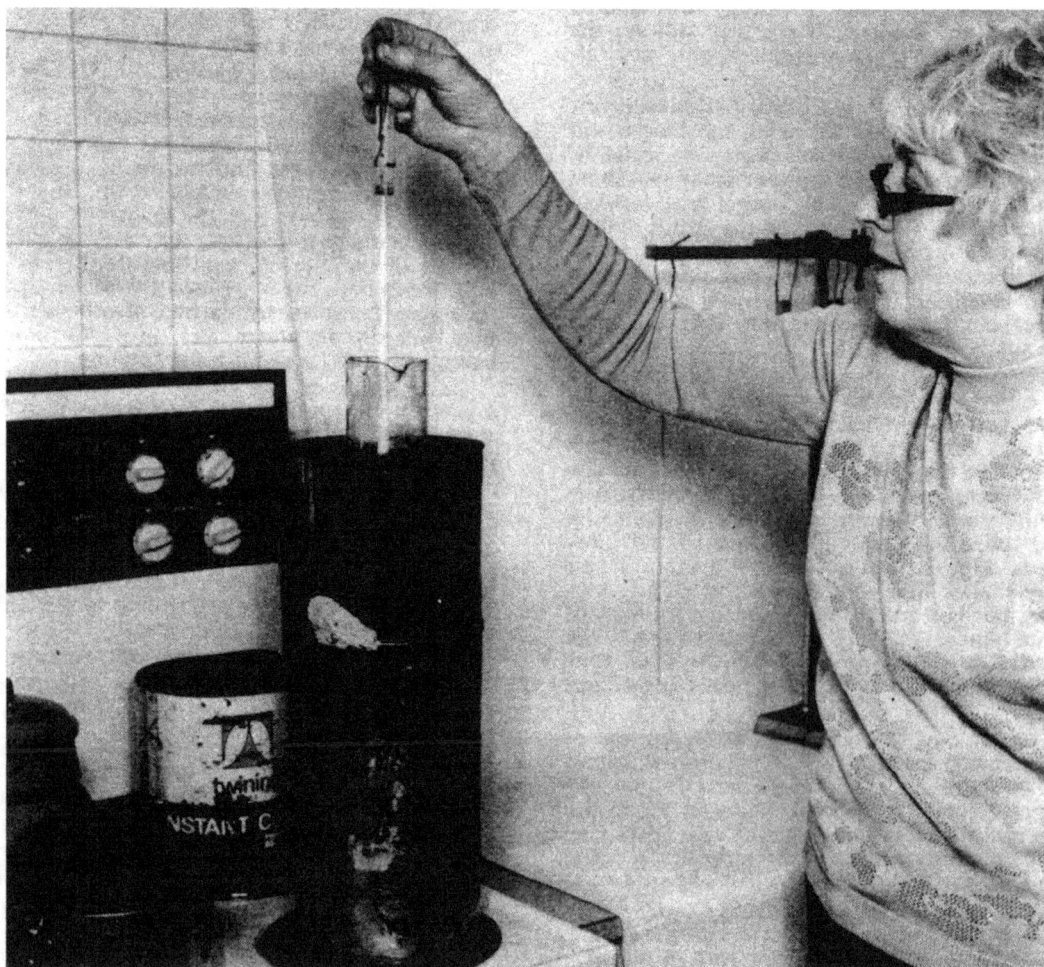

Apparatus used by author and described by her for dipped candles.

I made a dipping frame for four candles to fit into a gallon oil can (with top removed). A dipping frame, used in colonial America by the old settlers keeps the wicks rigid and

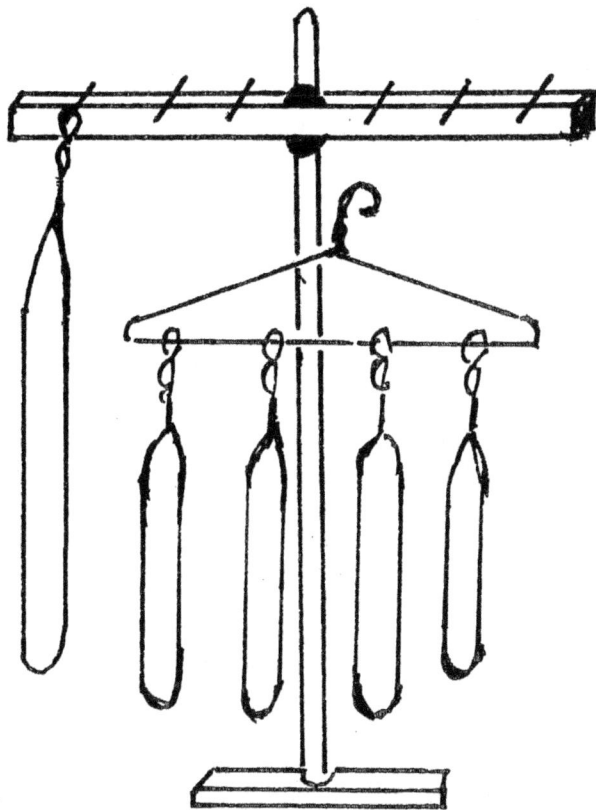

Two simple forms of candle stand: (a) wooden bar with nails supported by retort stand, (b) a wire coat-hanger.

obviates the necessity of repeated rolling. Unfortunately I never had enough high quality beeswax to fill it and so have only tested it out with wax-mixtures. It is effective but messy.

Melt your wax in a double pan and fill your dipping pot almost to the top. Stand your thermometer in the surrounding water and keep the whole on your hot plate. (Use asbestos pad if you have gas). Pour the wax in at 76°C (170°F) and wax your wicks, one at a time, by holding them in the wax until all bubbles stop rising. Hang to cool.

Fill some vessel as deep as your dipping pot with water at 38°-44°C (100°-110°F). Very soft (ideally distilled) water with a tiny bit of pure soap dissolved in it is ideal. Do not put soap into tap water.

Dip in one movement

Dip the wicks again, this time lowering them into the wax and drawing them out again in one movement, so that the pause at the bottom of the dip is only one to two seconds. When all have been dipped, roll each wick between sheets of glass or on any smooth surface with any other smooth sheet (glass, wood, plastic). If the wax tends to stick or flake, dip the candle in the water bath, wipe with a small, absorbent

towel and roll. You will gain experience on the best temperature, time and pressure for this operation.

Try to work without the water bath and you will make a better candle. Practice will teach you the best interval at which to roll – when your growing candle is neither too warm nor too cool.

Now your candle is quite straight. Dip again so that it touches the bottom of your cylinder. Two or three more dips and you will be able to see where the top of your candle is going to be. A dipped candle is right way up—the top of the candle tapers in a beautiful shape as the wax flows down the candle, setting after each dip. Roll after each alternate dip until the candle looks quite straight. If you are near a window or door and the candle tends to curve towards the cooler air, turn the peg at each dip to compensate for this. Try to achieve still air of an even temperature and your rolling will be much reduced.

Now reduce the temperature of your water bath to 69°-71°C (156°-162°F) and more wax will adhere at each dip.

As soon as the candle is about 5mm thick and is hanging straight, you can proceed with dipping and stop the regular rolling. Watch that you dip just as far as the chosen top of the candle to make a uniform tapering. An icicle will form on the bottom of the candle ("the port") and, in time, this will prevent your dipping the whole length of the candle. Cut this off with scissors or a sharp knife. As soon as you see any imperfection occurring in your candle, roll it again.

The elegant, tapering shape of your growing candle will not permit you to roll it with a straight rolling action. You will have to make curved sweeps with your rolling board. At this

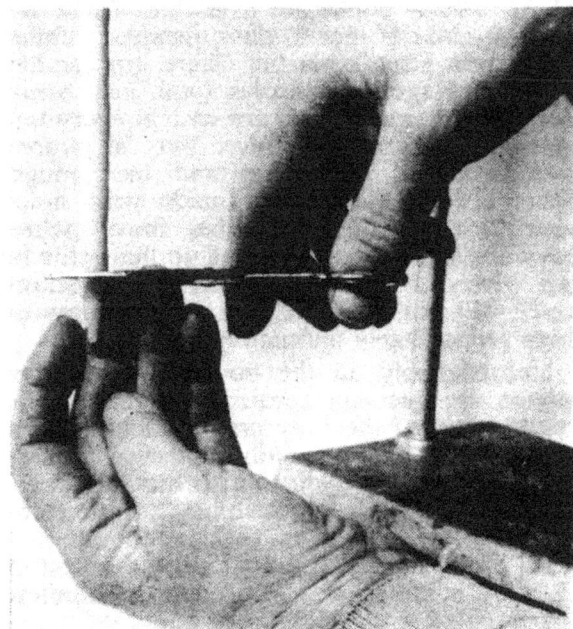

Trimming off the "port" or "icicle" which forms beyond the wick.

stage, I find it easier to use my hands. You will learn how much pressure is necessary to smooth, straighten and shape your candle and how to avoid the pressure which breaks it. A broken candle cannot be mended.

When the candle is as thick as you desire (i.e., it coincides with the wick you chose for its making), heat up the wax again several degrees more and do one or even two quick dips in the warmer wax to give a smooth finish.

Of course, you will realise that your dipping pot must be filled up from your double pan of melted wax at regular intervals. Take care not to over-fill, as the growing candle displaces an increasing amount of wax.

Measure the candles to ensure you have three exactly the same length, width and appearance. Flatten the base of each candle by holding it against the hot electric kettle or on the base of a gently heated frypan. If any defect appears on the tapering top, melt it away by deftly rolling it against the hot kettle.

Trim the wick to about 1cm ($\frac{1}{2}$ inch) and mature your candle for at least 24 hours before storing away.

If made for a specific candlestick, try your candle in it and make it fit snugly so that it remains upright without packing. Balance candle and holder to make a satisfying whole.

MOULDED CANDLES

All the American candle books refer to MOLDS and MOLDING. I often think this spelling is to be preferred to avoid confusion with "mouldy". However, here we come to the candles which are easiest to make, and which impress the uninitiated far more than the dipped or poured varieties which are the true craft of candle making.

There is wide interest in candle craft and a whole rash of books are available. There are moulds made of metal, glass, perspex, rubber and a new stretch plastic. There are candles made in disposable bottles (you just break them away) and candles are cast in every imaginable domestic container. Wax is poured into nutshells, wineglasses and beer mugs. There are gimmicky ones made with holes where incorporated ice cubes have melted away and wax is even whipped up like white of egg. Many of these candles have decorations which will not burn. In fact far too many of them are strangely unfunctional.

Unfortunately all the books tell you you cannot use beeswax because it sticks to the moulds. Indeed there are no moulded beeswax candles made commercially and even famous beekeepers will tell you that to mould a beeswax candle is "an impossibility". Even Root says that "though from the standpoint of a clear, bright flame, a beeswax candle is best of all, . . . it is the hardest to use and is definitely not practicable to mold".

The Modern Art of Candle Making (Olsen & Olsen) talks of the rich appearance and superior quality of beeswax, but says you cannot mould it. Those people who "are fortunate enough to have access to beeswax should mix it half and half with paraffin wax or just use 10% beeswax with 70% paraffin wax and 20% stearic acid". I have tried many combinations, but none has anything approaching the beauty of pure beeswax. Mixtures have only one advantage—they do not feel sticky. Neither does a beeswax candle when you have kept it a few months.

Anyone who has prepared beeswax for the show bench knows that it *can* be moulded, but that it is not easy. Those who have not, can learn by referring to the NHS Feature Article No. 3. "Wax for Show" by the late F. Padmore.

The secret of a good candle is to lubricate your mould with something which will prevent the wax sticking to the mould. Glycerine, liquid detergent or a mixture of the two are commonly used. Try also pure liquid soap. There must be total covering of the mould but not sufficient to form globules or bubbles which will ruin the finished candle. Paint into the pattern with a soft paint brush dipped in your liquid, then rub off any excess with a very soft piece of old well-washed cotton material —or with your clean hands. If you are using a rigid mould which cannot be turned inside out, you will have to poke your lubricant into the mould. A soft bottle brush with a piece of well-washed cotton tied round it is most effective.

A really effective mould release agent is a silicone spray. This modern preparation intended for lubrication of switches and tuners is most effective when used on rigid moulds – those made from metal or glass. In my experience, whilst it gives excellent results when sprayed into ornate rubber moulds, it does tend to shorten the life of the mould, making it less pliable. The blue plastic moulds require no mould release agent.

All moulds should be warm when you pour the wax. A fan heater or cool oven is safer than holding it over the cooker. Don't take risks with expensive moulds.

Most candle books recommend that pouring-wax should be about 11°C (20°F) warmer than its melting point. This temperature is recommended in candle factories where multiple moulds are in use. It applies to rigid moulds rather than to the ornate flexible type. This is to ensure sufficient contraction to draw the wax away from the mould as it cools.

In a large candle it also contracts away from the wick and a surprisingly large cavity can form. When the surface wax begins to set, but is still leathery, break it with a long skewer or needle and probe right into your candle. Take care not to dislodge the wick. Fill up the cavity with hot wax. You may have to do this several times whilst the candle is setting. Do not let any wax overflow the surface and run down between the candle and the mould. If you do you may not be able to release the candle and the surface will be spoiled.

If you have a really effective mould release, I find it preferable in the case of ornate candles to pour the wax when it is almost on the point of setting (65°-67°C, 150°-155°F). Wax holds heat better than the water in your water bath, so wait till a thin skin is beginning to form on the surface. Because there will be very little contraction when the wax sets, your design will have much more detail and precision. With

slow cooling you get no hollow developing inside the candle and little distortion. A slight hollow will appear on the surface. This may be filled carefully avoiding any overflow.

Slow cooling is essential. Candle craft books recommend rapid cooling—this is not for beeswax. Standing the mould in warm water ensures ideal cooling, but is fraught with risks. Mould and wax must be weighted to prevent floating and spilling. The water must come right up to the level of the wax or a ring will show on the finished candle. Any water splashed into the wax will ruin the pattern.

You will find effective alternatives. Wrap the mould in corrugated paper or newspaper. Suspend it in a china or glass jar. Stand it in a warm cupboard where it will be free from draughts, jarring or vibration.

Be sure your candle is quite cold before removing it from the mould. Leave at least overnight. A large candle may need even longer. Impatient candlemakers can be put off the craft for life by trying to extract a large candle when the centre is still liquid.

Let your candle mature in air for at least 24 hours, then wrap it to avoid damage. Beeswax candles lose their tacky surface if kept a few months.

Show candles must be identical. Blend sufficient wax to make at least four candles before moulding the first one. You can then choose the three you think most perfect and they will all be the same colour.

I look forward to a time when a number of flexible, embossed moulds can be made available to beekeeper-candlemakers in suitable designs.

CANDLE making at its best is like other crafts such as basketry—it cannot be perfectly reproduced by mechanisation.

To see the rings which indicate fifty dips on a large candle is like reading the age of a tree from its growth rings. The air introduced by this gradual building up of layers of wax makes for perfect combustion. Though traditions of craftsmanship such as this inspired the technical skill of the early industrialists, it is sad to see how few of the hand craftsmen are now retained. As John Dummelow says in his beautiful history of the Wax Chandlers, "Compression candles are really for new waxes and a new technology connected with it, but as the Church finds itself unable to afford pure beeswax, the proportion allowed (now 25%) can be translated into industrial methods". Several candle manufacturers told me this and that they had only recently suspended their hand-worker(s). So WE are likely to be the sole survivors—we beekeeper-candle-craftsmen—of a dying skill.

Having achieved our lovely candles, we must concede that they are to burn. Let us give our candles the best conditions for this. Candles must burn in quiet air. Even a well balanced candle may gutter or smoke in a draught or in otherwise agitated air. The storm lanterns of the past and the hurricane candles of the West Indian 'great-house' had beautiful protective glasses. People can have as many beeswax candles as they wish for a special festival but there will be no irritating fumes and no discoloration of the ceiling.

You will be surprised how long your beeswax candle will burn. In fact, far longer than almost all other candles. People ask, "How long?" Well, noble King Alfred (as all school children know) told the time by his beeswax-candle-clock. Take two identical candles, burn one for an hour and then mark the second one. Check after a second hour and then calculate its full life.

When you take your candles out of store, they may be marked by the characteristic bloom of beeswax. This is not a mould, nor does it betoken any deterioration in your candle. It is just the typical behaviour of beeswax, especially in the cold. Polish up your candles with an extra soft shoe brush or a piece of pure silk. Bloom melts at 38°C (102°F) so warming your candle a little, helps the polishing.

Beeswax candles are wonderful conversation pieces. Swot up a few anecdotes or facts about beeswax and about candles and your guests will be intrigued.

Do you know why "the game is not worth the candle" or how to "catch a thief in a candle"?

Of course too many of us burn the candle at both ends, but can we defend it by saying with the poet*

"*My candle burns at both ends;
It will not last the night;
But oh my foes and oh my friends—
It gives a lovely light*"

And . . . to conclude: What about getting together with our fellow mead makers, and, like the Worshipful Company of Wax Chandlers drink a loving cup at Candlemas and eat a special candlelit feast, asking this blessing:

"*For thy creature the bee
The Wax and the Honey
We thank Thee O Lord.
By the light of all men
Christ Jesus our King
May this food now be blessed.
Amen.*"

---

* Edna St. Vincent Millay, 1892-1950.

Candle Customs –

ANCIENT AND MODERN

IN SPITE of having immediate and almost excessive illumination at the flick of a switch, candles still burn in the modern world.

A candle lit in a tiny church outside Naples in 1921 should last until the year 3721. It is 18 feet high and one foot thick and was made as a memorial to Enrico Caruso, the great Italian tenor. It does not burn perpetually, but is lit once a year on the anniversary of his death. At this rate it should last 1800 years. I do not know what wax it is made from, but probably from "candlewax" – 80% paraffin and 20% stearin.

Mrs. May Horton wrote to me in December 1977, shortly before her death, telling me that her father remembered auction rooms in which pins were stuck in candles to determine the moment of sale. This was called "selling by inch of candle". The purchaser was the last bidder before the candle burned as far as the pin and caused it to fall. In the *Reading Mercury* for December 16, 1893, the Council is reported to have sold goods by this method, and they justified the choice by claiming it to be "the most probable means to procure the true value of the goods".

A similar system is still used annually in the sale of fine wines in aid of the Hospice de Beaune in the principal wine producing area of France. Local vintners present wines for sale by auction and wine buyers from all over the world attend. For each lot a starting price is decided by the administration of the Hospice, and a small candle is lighted at the moment of the first bid. This is called the "premiere feu" or "first flame". When the first flame is at the point of extinction, but the bids are still being made, the auctioneer may decide to light a second candle, called the "deuxieme feu". At the dying moment of the second flame the sale is made.

Sometimes a third candle is lighted, but in this event, this is called the second flame, and so the length of combustion is decided by the auctioneer in relation to the bids. He is an astute seller and knows his distinguished customers. He raises a large sum of money for the hospital and uses this ancient tradition of the tapers to crown the round of tastings, dinners, luncheons and junketings which make up the Three Glorious Days devoted to Burgundy wines in Beaune every November.

There are several candle manufacturers in the U.K., and it is surprising how many candles are sold. Price's make about a million candles a day, many of them coloured in jewel colours. Red is the favourite. I guess this is because Christmas is the time at which most people dine by candle light. We are a safety conscious nation, and our Christmas trees no longer carry the real candle in its metal holder. We have substituted the electric fairy light. But we export thousands of Christmas tree candles to other nations which keep the tradition. Our Christmas candles are on our dining tables and on the thousands of Christmas cards which feature the familiar flame.

In 1962 I saw a report in the *American Bee Journal* of the "Candle Teas" held by the Moravian Church in Winston Salem, North Carolina. I wrote to the Church in 1978 and received a full report of that year's festivities. When the Germanic, Moravian immigrants founded Salem in November 1753, they were a pastoral community. Beekeeping was a thriving prospect for the Brothers, and there was wax in sufficient quantity for the making of pure beeswax candles for many years. The first candle service noted in Moravian records was held in Europe in 1747, when lighted candles tied with red ribbon were given to the children. Nowadays in the New World, industry has supplanted the farming activities, and the wax used for the candles is a mixture of four parts of tallow to one part of beeswax. They are moulded in replicas of the original Moravian tin moulds, using tobacco twine for wick. Each candle is decorated with a fire-proofed red paper frill to prevent hot wax falling on to the hand. The Christmas Eve Service is called the "Love Feast". Each participant carries one of the specially made candles and receives a programme of hymns and a short history of the customs.

In addition, Candle Teas are held during the first two weeks of Advent. These have become so popular, that it is not unusual for lines of guests to wait over an hour to gain admission. The Women's Fellowship of the church open their doors at two o'clock and until nine demonstrate the making of the candles, and serve coffee and a simple sugar cake. They dress in the traditional costume of their eighteenth century ancestors. From small beginnings, a popular tradition has gained much publicity. Nowadays thousands of candles are sold, many to other Moravian communities. The money goes to support a home for the aged and missionary projects.

Here in England many churches are reviving the Christingle service on similar lines. As in Salem, the candle, representing the light of the world, is tied with a red ribbon, symbolic of redemption. Children are given an orange in which to stick the candle and other fruits are added. This is said to represent the round world and God's bounty.

We have all seen the Russian dancers who execute the rapid, smooth footwork of their traditional folk dance with lighted candles on their heads. These commemorate the custom of lighting such candles to celebrate the return of Springtime after the dark Winter. Such celebrations are part of the folklore of all Northern peoples.

Candle festivals without old tradition are being introduced even today. At the Bath festival, townsfolk are asked to place lighted candles in their windows to attract the attention of the many tourists. When I asked the organisers what this commemorated, I was told there is no historical precedent, but that the committee thought it would be attractive. Other towns with music, drama or historical festivals use torches and candles to add glamour to the proceedings.

In November 1983, I saw in the *Evening Argus* a photograph of children at the Thai house of the Pestalozzi International Children's Village lighting candles for their "Night of Lights" ceremony. The "Loy Krathong", to give it its Thai name, takes place on the full moon night of the twelfth month of the lunar year. In the old days the candles were made by hand of beeswax, but recently any candle is used. Small boats, made from strips of banana leaf made into a cup shape, are trimmed with flowers; and into these are stuck a candle and three incense sticks. The rising moon is the signal for the launching of these miniature boats from the steps of homes, the sides of boats, the banks of every canal, river and lake. Candle flames flicker and glow and the fragrance of incense steals along the breeze. As each small flower boat is set adrift, a prayer is said: "With this Krathong we pay homage to the footsteps of Lord Buddha. May it result in our happiness and render us assistance for ever." If your candle stays alight as far as you can see, the wish will be granted.

In several wine festivals of the Rhine and its tributaries, the celebrations end with fireworks on the banks and on the stream a cavalcade of tiny plastic floats, each one holding a short, stubby candle. The light glows through the transparent holder, and the river's surface dances with hundreds of these flickering flames. It is a splendid sight.

And, of course, there are birthdays. No anniversary would be complete without its candle bearing cake. The lovely picture of the baby, eyes shining in wonder at the flickering sign of the passing year, is a feature of every photograph album. As the years progress, we still enjoy the big blow to the tune of "Happy birthday to you". When I demonstrate candle making at shows, I often let the children dip their own birthday candles. No candle, however brightly coloured, can match the natural beauty of pure beeswax.

Some say that candle magic starts with birthday cakes, a survival of ancient candle ritual*. Colours of candles have, for them, mystical significance: white or red for love, gold for power, blue for contentment, green for growth or wealth, black for hatred. There are candles to be lighted at the significant moments of birth, marriage and death, to drive off evil spirits which delight to work in darkness. Strange rituals with occult symbolism, zodiacal signs and magical formulae are designed for the superstitious amongst us. I am not one of them. I just want to encourage other beekeepers to keep the fine, old traditional craft alive by having a go at candle making. Let the light of beeswax keep on shining.

Bibliography. VINCI, LEO., *The Book of Practical Candle Magic.*

JUDGING BEESWAX CANDLES

THE show season is upon us again. There is a judge's handbook to remind the "experts" of the recommended techniques for assessing the merits of honey, wax and mead. In the case of cakes, sweetmeats and candles, the professional or commercial expert is often invited to judge. There are, however, many small shows where funds run to only one judge who tries to be jack-of-all-trades.

In my heart, I much prefer the instructive display to the competition bench. But, since competitive shows are still widespread, I offer a few suggestions to judges of the candle class. It is well nigh impossible to compare the merits of a candle made by pouring molten wax into a purchased mould with those of a handmade candle. These may be built up gradually by casting molten wax down a wick, dipping the wick into the wax, or both. They may be formed from sheets or strips of wax. The craftsman candle would always take precedence, so I hope there will always be two distinct classes.

GENERAL NOTES

The schedule usually asks for two or three candles, of which one will be burned. It expects identical candles (and should say so). They should be made from one batch of wax.

If the candles are not identical (or as near identical as is possible) there is no point in burning only one of them. The one burned should imply the same or similar performance in the other(s). Since the points looked for in a candle are so various and since the entries vary far more than entries in any other class, judging by elimination is impracticable. A points system is indicated.

Candles are to burn and this should be borne in mind when entering the candles in any show. They should be eminently functional. Ornate casts with little symmetry which cannot sustain a consistent flame have no place on the show bench. The main factors going into a good flame are: clear wax, good wick, consistent shape. The balance between size of candle and wick is vital. We produce our own wax, nevertheless, it would be most useful if a supply of good wick, specially pickled for beeswax in a large range of sizes could be made available to beekeepers at a reasonable price by the appliance trade.

Why must a judge extinguish a candle and relight it after cooling? A candle is never burned right through, it is extinguished and relighted many times during its "lifetime". It is essential that it continues to light easily and burn consistently until almost fully consumed.

JUDGING POINTS
Appearance of candle(s)
a) **Colour**

A light wax is preferred, but a darker wax should not be penalised if it is pleasing to the eye and not dull.

b) **Size**

The candle should be of a reasonable size to be judged in one hour's burning, as that is the maximum time available to most judges. Nevertheless a tall, fat candle is to be preferred to a small, thin one. In the case of moulded candles, the tip should be consumed within the hour, so that the regular column of the candle can be judged in performance.

c) **Symmetry**

Whatever the design of the candle, it should be generally symmetrical. There should be a good tip.

d) **Feel**

Beeswax is slightly tacky—feel can indicate absence of adulteration by other waxes.

Surface should be smooth and well finished. In the case of embossed or decorated candles, pattern should be definite and regular.

e) **Wick**

It should be of correct material and plaited.

The wick should be central at both top and bottom of the candle. Where the candle has a tip, the wick should project from the tip and not at one side. A wick should be trimmed to a suitable length.

f) Cleanliness

Wax should not have picked up any surface soiling.

Wax should be carefully cleaned. In the case of moulded candles, small specks of soiling will fall to bottom of mould, which is the top of the candle and will be visible in the tip.

The true cleanliness of the wax will not be apparent until the candle has burned for a time. Soiling in the wax will then be visible in the pool of molten wax which forms underneath the flame. In extreme cases, soiling will clog the wick and the flame will splutter.

[Since presentation is a factor in the exhibition of honey, mead, etc. it seems desirable that candles should be presented as a pleasing whole. Care in presentation can be evidenced by choosing a candlestick or holder to give a well balanced appearance in length, width, colour, etc. There is a danger that an expensive candlestick or lavish holder would influence judgement. This must be avoided. If this factor is to be considered, this should be indicated in the schedule.]

BURNING PERFORMANCE

A judge should not light a candle unless he or she is willing to observe its burning for an hour. A candle maker can reasonably protest if his exhibit is destroyed by being lighted, but not burned long enough to demostrate its full performance. Choose a place with no draughts.

a) Initial lighting

Should be easy. The wick should be long enough to catch the flame, but not so long that the wick falls into the melting wax.

b) Flame size

Should be in relation to candle size. This indicates balance of chosen wick to diameter of candle.

c) Flame shape

Should be rounded with very slight feathering at the tip. Beeswax tends to smoke a very little and this divides the tip of the flame into feathers. There is only a slight hint of this.

d) Brightness

If wick and candle are well balanced, the correct amount of molten wax is drawn up the wick and an optimum brightness of light for that wick is achieved.

e) Pool of Wax

This should almost fill the diameter of candle. Molten wax is transparent and reveals any soiling in the wax. Dirty wax will inhibit size and brightness of flame. If the wick is too thick it will draw up too much wax and the pool will be inadequate. If the wick is too thin, it will not consume sufficient wax and the pool will overflow.

The flame should be central to the pool of wax. This will naturally occur in a rolled, dipped or cast candle. In a moulded candle, if the wick is not central, the wall of wax on the outer edge of candle will be thick at one side and thin on the other. In time the thin side will burst and the pool of wax flow away or the thick side will project upwards towards the flame and cause uneven burning.

Very slight guttering is acceptable, especially in poor conditions of most halls, but excessive guttering should be penalised.

Wick Curve

This is the most important observation. The curve should be a gentle right angle and the burned tip should be continually and consistently consumed in the hot outer edge of the flame. The wick should not curl up like a pig's tail or growing fern, or it will curl into the pool of wax and cause excessive guttering. It may even drown.

EXTINGUISHING and RELIGHTING

The flame should be extinguished with a snuffer. If you blow out the flame, practise carefully to tongue the expiration like a trumpet player. Breath should not reach pool of wax, but just point of flame.

Observe the afterglow. The glow should be out in a maximum of 20 seconds, but 10 seconds is ideal. If the tip of the wick has an accumulation of carbon (known in the trade as a cauliflower) it will glow for a long time. In extreme cases it may fall into the pool of wax and cause subsequent severe guttering. Long afterglow may consume so much wick as to make relighting difficult. Some unsuitable (usually un-pickled) wicks may remain hot and collapse into the molten wax, become flooded and not stand up again.

Allow at least $\frac{1}{4}$ hour for cooling—so that the pool of wax solidifies and the second lighting can be properly observed.

Size, brightness and burning-off of the wick should be fully observed again after second lighting.

BIBLIOGRAPHY

BENNETT, H., *Industrial Waxes*, Vol. II.

BULL, Dr. Phil. R., *Vom Wachs*.

CLARK, Mrs. R. E., Notes taken at a lecture.

COWAN, Wm., Article in *Bee Craft*, Nov. 1946 (unsigned).

CRIBB, A. L., Article in *Bee Craft*, Dec. 1944.

CROSS, *Oxford Dictionary of the Christian Church*.

DUMMELOW, J., *The Wax Chandlers of London*. Phillimore 1973.

GROUT, R. A. (Ed.), *The Hive and the Honeybee*.

KNAGGS, N. S., *Adventures in Man's First Plastic*.

OLSEN & OLSEN, *Modern Art of Candle Creating*.

PURCHAS, S., *A Theatre of Politicall Flying Insects*. 1659.

ROOT, A. E., *Beeswax*, 1951.

STURDY, H. A., "WAX". A lecture published by Central Association of Beekeepers, 1972.

TAYLOR, R., *Beeswax Molding and Candle Making*.

VINCI, Leo, *The Book of Practical Candle Magic*, 1981.

WEISKE, W., *Die Entwicklung der Kerze*.

WILSON, G. F., "The Stearic Candle Manufacture". A lecture given at R.S.A. 1852.

www.ingramcontent.com/pod-product-compliance
Lightning Source LLC
LaVergne TN
LVHW080322090426
835511LV00037B/1926